an·tag·o·nist *n.* opponent or adversary —**an·tag·o·nis'tic** *adj.*
an·tag·o·nize' *v.* (-nized, -niz·ing) make hostile; provoke
An·tal·ya /äntäl'yə/ *n.* seaport in Turkey. Pop. 378,200
An·ta·na·na·ri·vo /an'tanan'ərē'vō/ *n.* capital of Madagascar. Pop. 802,400
Ant·arc·ti·ca /antärk'tikə, ärt'ikə/ *n.* 1 ice-covered continent surrounding the South Pole 2 (the Antarctic) region of Antarctica and surrounding oceans —**ant·arc'tic** *adj.* [L fr. Gk, rel. to ARCTIC]
Ant·arc'tic Cir'cle *n.* parallel of latitude 66° 32' S, delimiting the Antarctic region
an·te /an'tē/ *n.* 1 stake put up by a player in poker, etc., before receiving cards 2 amount payable in advance —*v.* (-ted or -teed, -teing) 3 put up as an ante
ante- *prefix* before; preceding [L: before]
ant'eat'er *n.* any of various mammals feeding on ants and termites
an·te·bel·lum /an'tēbel'əm/ *adj.* before the American Civil War (1861–65) [L]
an·te·ced·ent /an'təsēd'nt/ *n.* 1 preceding thing or circumstance 2 *Gram.* phrase, etc., to which another word or ative pronoun) refers 3 (*pl.*) history or ancestors —*adj.* ANTE- + *cedere* go]
an'te·cham'ber /an'tē-/
an'te·date' /an'tē-/ *v.* precede in time 2 as
an·te·di·lu·vi·an /an'tədə the time before the or out of date [L
an'te·lope /an'l like ruminant, e.g. [MGk *antholops*]
an·ten·na /anten each of a pair of of wires, rods, et sects, crustacean of wires, rods, et radio waves [L:
an·te·ri·or /antēr' 2 prior [L fr. *ante b*
an'te·room /an'tiroom ing to a main one
an'them /an'THəm/ ture 2 solemn hymn of prais tional anthem [L fr. Gk, rel. to
an'ther /an'THər/ *n.* part of a stam ing pollen [Gk *antheros* flowering]
ant'hill' *n.* moundlike nest built by ants or termites
an·thol·o·gy /anTHäl'əjē/ *n.* (*pl.* -gies) collection of poems, essays, stories, etc. —**an·thol'o·gist** *n.*; **an·thol'o·gize** (-gized, -giz·ing) *v.* [Gk *anthos* flower + *-logia* collection]
An·tho·ny /an'THənē/, Susan B(rownell) 1820–1906; US suffragist
an·thra·cite' /an'THrəsīt'/ *n.* type of hard coal burning with little flame and smoke [Gk, rel. to ANTHRAX]
an·thrax /an'THraks/ *n.* disease of sheep and cattle transmissible to humans [Gk: coal; carbuncle]
an·thro·po·cen·tric /an'THrəpəsen'trik/ *adj.* regarding mankind as the focal point of existence [Gk *anthrōpos* man]

23   antagonist / antiknock

an·thro·poid /an'THrəpoid'/ *adj.* 1 human in form 2 apelike —*n.* 3 anthropoid ape
an·thro·pol·o·gy /an'THrəpäl'əjē/ *n.* the study of mankind, esp. its societies and customs —**an'thro·po·log'i·cal** /-pəläj'ikəl/ *adj.*; **an'thro·po·log'i·cal·ly** *adv.*; **an·thro·pol'o·gist** *n.*
an·thro·po·mor·phism /an'THrəpəmòr'fiz'əm/ *n.* attribution of human characteristics to a god, animal, or thing —**an'thro·po·mor'phic** *adj.*; **an'thro·po·mor'phi·cal·ly** *adv.* [Gk *anthrōpos* man + *morphē* form]
an·thro·po·mor·phous /an'THrəpəmòr'fəs/ *adj.* human in form
an·ti /an'tē, an'tī/ *prep.* 1 opposed to —*n.* (*pl.* -tis) 2 person opposed to a policy, etc.
anti- *prefix* 1 opposed to (*anticlerical*) 2 preventing (*antifreeze*) 3 opposite of (*antithesis*) [Gk: against]
an'ti·a·bor'tion *adj.* opposing abortion —**an'ti·a·bor'tion·ist** *n.*
an'ti·air'craft *adj.* (of a gun or missile) for use to attack enemy aircraft
-tic /an'tēbī-ät'ik, an'tī-, -bē-/ *n.* 1 that can inhibit or destroy suscep- organisms —*adj.* 2 functioning as [Gk *bios* life]
(*pl.* -ies) a blood protein pro- onse to and then counteracting
(usu. *pl.*) foolish behavior ico ANTIQUE]
tikrīst'/ *n.* enemy of Christ /an'tikrīs'CHən, an'tī-/
əpāt'/ *v.* (-pat·ed, -pat- use before the proper time 2 forestall (a person or to —**an·tic'i·pa'tion** [L *anti-* before + *cap*
naks', an'tī-/ *n.* dis- clusion to something ac'tic /-klīmak'tik/ -ly *adv.*
idepres'ant, an'tī-/ *n.* depression —*adj.* 2 al- on
tidōt'/ *n.* 1 medicine, etc., ounteract poison 2 anything con- ng something unpleasant [Gk *antidotos* ven against]
an·ti·freeze' /an'tifrēz', an'tī-/ *n.* substance added to water to lower its freezing point, esp. in a vehicle's radiator
an·ti·gen /an'tijən/ *n.* foreign substance that causes the body to produce antibodies [Gk *-genēs* produced]
An·ti·gua and Bar·bu·da /antē'gwə; bärbōōd'ə/ *n.* island state in the West Indies. Capital: St. John's. Pop. 64,000
an·ti·he·ro /an'tē hēr'ō/ *n.* (*pl.* -roes) (in a story) central character lacking conventional heroic qualities
an·ti·his·ta·mine /an'tihis'təmēn', an'tī-, -mən/ *n.* drug used in treating allergies
an·ti·knock /an'tē·tinäk', an'tī-/ *n.* substance added to motor fuel to eliminate knocking noise produced by premature combustion

anthracite one.

Copyright ©2002 **anthracite**.

All rights reserved. No part of this book, including interior design, interior images, cover design, and specific design elements, may be reproduced or transmitted in any form, by any means (electronic, photocopying, recording, or otherwise) without the prior written permission of the publishers.

Designed and written by members of **anthracite** unless otherwise identified.

anthracite is Rick Blasdell, Russ Cox, Gene Hosey, Jeffrey King, and Jerry King Musser.

First printing June 2002. Edition of 500.

Printed and bound in the United States of America.

ISBN 0-9719456-0-8

an·tag·o·nist *n.* opponent or adversary —an·tag·o·nis'tic *adj.*
an·tag·o·nize' *v.* (·nized, ·niz·ing) make hostile; provoke
An·tal·ya /äntäl'yə/ *n.* seaport in Turkey. Pop. 378,200
An·ta·na·na·ri·vo /än'tənan'ərē'vō/ *n.* capital of Madagascar. Pop. 802,400
Ant·arc·ti·ca /äntärk'tikə, ·ärt'ikə/ *n.* 1 ice-covered continent surrounding the South Pole 2 (the Antarctic) region of Antarctica and surrounding oceans —ant·arc'tic *adj.* [L fr. Gk, rel. to ARCTIC]
Ant·arc'tic Cir'cle *n.* parallel of latitude 66° 32' S, delimiting the Antarctic region
an·te /än'tē/ *n.* 1 stake put up by a player in poker, etc., before receiving cards 2 amount payable in advance —*v.* (·ted or ·teed, ·teing) 3 put up as an ante
ante- *prefix* before; preceding [L: before]
ant'eat·er *n.* any of various mammals feeding on ants and termites
an·te·bel·lum /än'tēbel'əm/ *adj.* before the American Civil War (1861–65) [L]
an·te·ced·ent /än'təsēd'nt/ *n.* 1 preceding thing or circumstance 2 *Gram.* phrase, etc., to which another word [...] ative pronoun) refers 3 (*pl.*) [...] history or ancestors —*adj.* [...] ANTE- + *cedere* go]
an'te·cham'ber /än'tē- [...]
an'te·date' /än'tē-/ *v.* 1 [...] precede in time 2 ass[...]
an·te·di·lu·vi·an /än'tē[...] the time before the [...] or out of date [L A[...]
an·te·lope /än'təlō[...] like ruminant, e.g. [...] [MGk *antholops*]
an·ten·na /änten'[...] each of a pair of [...] sects, crustacean [...] of wires, rods, et[...] radio waves [L: s[...]
an·te·ri·or /äntēr'[...] 2 prior [L fr. *ante* b[...]
an·te·room /än'tirōom/[...] ing to a main one
an'them /än'thəm/ [...] composition usu. based on [...] ture 2 solemn hymn of prais[...] tional anthem [L fr. Gk, rel. to [...]
an·ther /än'thər/ *n.* part of a sta[...] ing pollen [Gk *anthēros* flowering]
ant'hill' *n.* moundlike nest built by ants or termites
an·thol·o·gy /änthäl'əjē/ *n.* (*pl.* ·gies) collection of poems, essays, stories, etc. —an·thol'o·gist *n.*; an·thol'o·gize (·gized, ·gizing) *v.* [Gk *anthos* flower + *-logia* collection]
An·tho·ny /än'thənē/, Susan B(rownell) 1820–1906; US suffragist
an·thra·cite /än'thrəsīt'/ *n.* type of hard coal burning with little flame and smoke [Gk, rel. to ANTHRAX]
an·thrax /än'thraks/ *n.* disease of sheep and cattle transmissible to humans [Gk: coal; carbuncle]
an·thro·po·cen·tric /än'thrəpəsen'trik/ *adj.* regarding mankind as the focal point of existence [Gk *anthrōpos* man]

23 antagonist / antiknock

an·thro·poid /än'thrəpoid'/ *adj.* 1 human in form 2 apelike —*n.* 3 anthropoid ape
an·thro·pol·o·gy /än'thrəpäl'əjē/ *n.* the study of mankind, esp. its societies and customs —an'thro·po·log'i·cal /-pəläj'ikəl/ *adj.*; an'thro·po·log'i·cal·ly *adv.*; an'thro·pol'o·gist *n.*
an·thro·po·mor·phism /än'thrəpəmôr'fiz'əm/ *n.* attribution of human characteristics to a god, animal, or thing —an'thro·po·mor'phic *adj.*; an'thro·po·mor'phi·cal·ly *adv.* [Gk *anthrōpos* man + *morphē* form]
an·thro·po·mor·phous /än'thrəpəmôr'fəs/ *adj.* human in form
an·ti /än'tē, an'tī'/ *prep.* 1 opposed to —*n.* (*pl.* ·tis) 2 person opposed to a policy, etc.
anti- *prefix* 1 opposed to (*anticlerical*) 2 preventing (*antifreeze*) 3 opposite of (*antithesis*) [Gk: against]
an'ti·a·bor'tion *adj.* opposing abortion —an'ti·a·bor'tion·ist *n.*
an'ti·air'craft *adj.* (of a gun or missile) for use to attack enemy aircraft
an'ti·bi·ot'ic /än'tēbī·ät'ik, an'tī-, ·bē-/ *n.* 1 [...] that can inhibit or destroy susceptible organisms —*adj.* 2 functioning as [...]
an'ti·bod'y *n.* (*pl.* ·ies) a blood protein pro[...] to and then counteracting [...]
[...]ity (usu. *pl.*) foolish behavior end 1[...] *ico* ANTIQUE]
• Usage: [...]
an'ti·christ' /[...]tikrīst'/ *n.* enemy of Christ [an·ti·chris·tian /än'tikris'CHən, an'tī-/ [...]
[...]o·pat'/ *v.* (·pat·ed, ·pat[...] use before the proper time [...] forestall (a person or [...] to —an·tic'i·pa'tion [...] [L *anti-* before + *cap[...]
[...]aks', an'tī-/ *n.* dis[...]clusion to something [...]mac'tic /-klīmak'tik/ [...]l·ly *adv.*
[...]idipres'ənt, an'tī-/ *n.* [...] depression —*adj.* 2 al[...]
[...]tidōt'/ *n.* 1 medicine, etc., [...]counteract poison 2 anything countering something unpleasant [Gk *antidotos* given against]
an·ti·freeze /än'tifrēz', an'tī-/ *n.* substance added to water to lower its freezing point, esp. in a vehicle's radiator
an·ti·gen /än'tijən/ *n.* foreign substance that causes the body to produce antibodies [Gk *-genēs* produced]
An·ti·gua and Bar·bu·da /äntē'gwə; bärbood'ə/ *n.* island state in the West Indies. Capital: St. John's. Pop. 64,000
an·ti·he·ro /än'tē hēr'ō/ *n.* (*pl.* ·roes) (in a story) central character lacking conventional heroic qualities
an·ti·his·ta·mine /än'tihis'təmēn', an'tī-, ·man/ *n.* drug used in treating allergies
an·ti·knock /än'tinäk', an'tī-/ *n.* substance added to motor fuel to eliminate knocking noise produced by premature combustion

Sun & moon knife
night & day
into neat LITTLE edible slices//
What//is suspended
over what was/ **superimposed//**

it never glitters
still it's up against a duller color:
we glimpse a bit
of vast impermanence
but feel the tug of leash:

like gold

a teacup full of

lucky leaves
or verses full of daggers /
a magic act
where the magician waits
with his casket face
smile smeared on
eyes half-sewn shut//

...a child
small & smiling
    walks an empty quiet street
    toward another
    small & smiling
    with a tail
    and hoofs for feet...
    now you see it / now
    it sees you
      & doesn't like what it's
        going to be ...

GO!! GO NOW!!
ORDER DAMNIT!
ORDER IN THIS UNIVERSE!!!

scrap of wood
with fish-eye knothole
dances in flames
surfaces
        &swims   a   w   a   y
behind curtains of fire
coming **ALIVE**
in the flick of the fire that consumes it
   like pages of books
     in fevered imaginations/
       or a poem//read out loud:
         the fish swims
            when the poem's read

                                  out loud

Funny how you can

think like that/

THE geometry of **MEMORY** isolation

relying on//something **THAT** isn't there

FORGETTING for a moment it's missing

thinkin it's there for the askin

when//whatcher askin really **IS**

which is the weak one in the herd?/

find the **WEAK** one & lunge ....

It never occurred to any of us
that Superman was ever merely
an arrangement of lines on Paper

In the beginning

is the tricking of the senses /

the tap-tap-tapping of the 6th sense

pulling us thru the window

We're taught not to stare
as children / and gravity
is the motion of a coin ,
to the bottom of a pocket and
..it's better to stay indoors
and not to stare at living things

   ju_t glance/& look away so you can
   give the lines their motion
   And you build new things
   with the things you hear
   run yer hands over a
      neat new language
that whispers thru this vastness
    where humans become

something like rumors]
swirling thru circuits in cyphers]
raising zero-tolerance zombies
who will trudge along / in their father's shoes
one day soon enough / digging in their heels

digging in their pockets

for those steady falling coins /

Pleasant & saying thank you for having

the privilege of "buying whatever they want

and then / buying the things that will kill them

the 6th sense tappeth.
    and Superman flies /

Clark Kent tears off

his standard-issue

real-world suit

stripping off his necktie first to gulp in

panicked oxygen

borders of the cartoon panels

go at first translucent

as the black lines swell

with Clark Kent's inrush of

button-popping air and till

the tie & suit & shoes are gone

Superman flies and I fly

thru the pages / thru his eyes

rocking in a tree where
'no one will ever see me / no one will ever find her here
no one will ever think of it'

cause nobody ever looks up

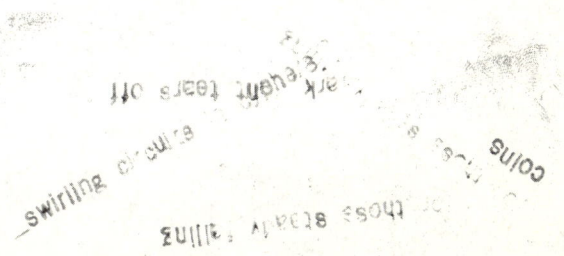

Why Can't I Write?

Why? Nothing to say. Lack of education or just too stoopid. Too many words, so little typiing skills. My pens keep breaking. Ink blotting like a Rorschach test. I don't have the answer. All of the above. None of the above. One of the above. The question squeaks like a hinge in my mind .. NO! YES! No. A lock. A deadbolt. Nah. Squeaky like a hinge. YES! YES! YES! LOUD and distracting. Or rusty. A rusty, squeaky hinge. And old. OLD!?!? Now I did it.

I see, well, read, stare, fume over people's prose and musings all the time. I say to myself, "where did the words come from?" Thoughts. Words dancing with the grace of Gene Kelly. It amazes me to no end nor any beginning. It is as if they have become one with their computer. A mighty zen. Zen of the pen.

My friends write. Or at least attempt. They're visual. Artist types. They write short stories, essays, and screen plays. Channeling into their nimble fingers. Abusing mother nature's herbal means. Who said that? Not I. Just read what they have comitted to paper. Bradbury would blush.

People tell me I have plenty to say. So did Meher Baba. I did try as a kid. Spending my early years having to "sit down and shut up". Maybe that has put this deadbolt around my writing craft. Yeah, a craft. Which craft. YEAH WITCH CRAFT! That's it you bastards! Like a Salem witch who dabbles, pours, stirs and rants. There is a magic to it. There was a time after my "sit down and shut up" years that I attempted to write. My arms got very tired from using the Comet and rag to wipe clear my iambic pentabulator. Shit! Where did that word come from. Mrs. Johnson, wherever you are, I finally used a word that you taught me. Big whoop. She would be happy if I spoke in complete sentences. Where was I? On some southbound train of thought.

I just dunno. It could be that my Southern Roots will show. My identity will be exposed as if Clark Kent left his glasses in the outhouse. You see, being a Southern lad, (lad?) there isn't a history of writers in my family stump. I was the first to go to college. Okay, okay, okay. Art school ain't no college. And yes, I chose it becuase of the lack of English, Math, etc. but that ain't da point. What point? West point. There isn't a histry of righters in my bloodline. Inbredline. None of my fambly signed dem dar Declaration of Independences. Many of my kin folks probly could not sign their dang burn name at the time. Couldn't scratch an "x" into a log today. Yoons just gotta understand. It ain't dat important. Moonshine, fishin', tillin' and yankin' weeds. That's the life. But I got my penmanship.

Real purdy like.

I do want to write something. Anything. One thing. Emails do not count. Oh hell. Let's not get greedy. Needy. I just can't find the words. Thoughts. Verbage. And storyline. I can draw a right purdy picture, but why can't I write.

The nightmares of a catholic child are built in vast dark moving spaces & the spac
 The nightmares of a catholic child growing up in the pennsylva
 The human race in general, the individuals who have survived, stand at the ed
 Wide chasms of opportunity yawn just past the tips of our wiggli
 With a single leap of faith one can transport oneself like a joyous bird o
 One day follows another, like footprints filling with water and disappeari
 My own first stuttering footsteps kicked up the radioactive anthracite coal d
 This dust still surrounds the holes that keep opening up in front of r
 No diamonds twinkle from the potholes on my path. The potholes just drop open l
 The next trap d
 The glowing dust at the edges of crashed-in mines.
 Crashed in like bomb crat
 Sometime between the 40's & 50's as the mines played out, the owners reluctan
 whole era. All their industry, all their battles, all their bandaged lim
 An entire generation goi
 When the trap door opens you drop out.
 You drop in.
 You abruptly fall. You're pushed. You want

However it happens, you're history.
But only one molecule in an atom of
 a history of one species on a planet
 of millions. One planet in a universe
  of billions. A shade of scattered
 glowing dust on a pile of rubble in
 the fiction of a race that's existed for
 one-thousandth of the width of a
   human hair at the end of 300 feet
 of time's oily rope.

"see you

   Traces are everywhere.

   Traces of everything.

   Everything but money.

Welcome to the mines.

m with vivid hellish images torn from the tortured souls of tormented nuns & priests.

l regions ten years after the coal runs out are more immediate.
a precipice of bottomless promise.

s – vague promises glint from the black quilt spread before us.
cathedral spires of global suckcess, chimes jangling glory hallelujah.

pped at the end of one war into the middle of the next.

ored darkness, a yawn fulla coal hinged halfway between my toes and the horizon.

t crashes open could swallow me but no one down here even notices me.
Trapped in the dust and the dust trapped in us.

a few nickels more.

sed them. Like snapping light switches off, darkening windows one at a time on their
ry serious matter, every outburst of laughter, disappears without a trace.

bed.   Well, not without a trace.

're enticed.   Some even jump.

later, navigator."

Birth & a lifelong struggle with gravity

He thrashed in a sea of pure ideas

are not becoming to

a world's highest life form/

Perhaps the picture was too perfect

and sounds / like pitched stones

began to ripple the image

stirring up mud / and / perhaps

the mud looked like a memory

memory like a sound

and sounds became words

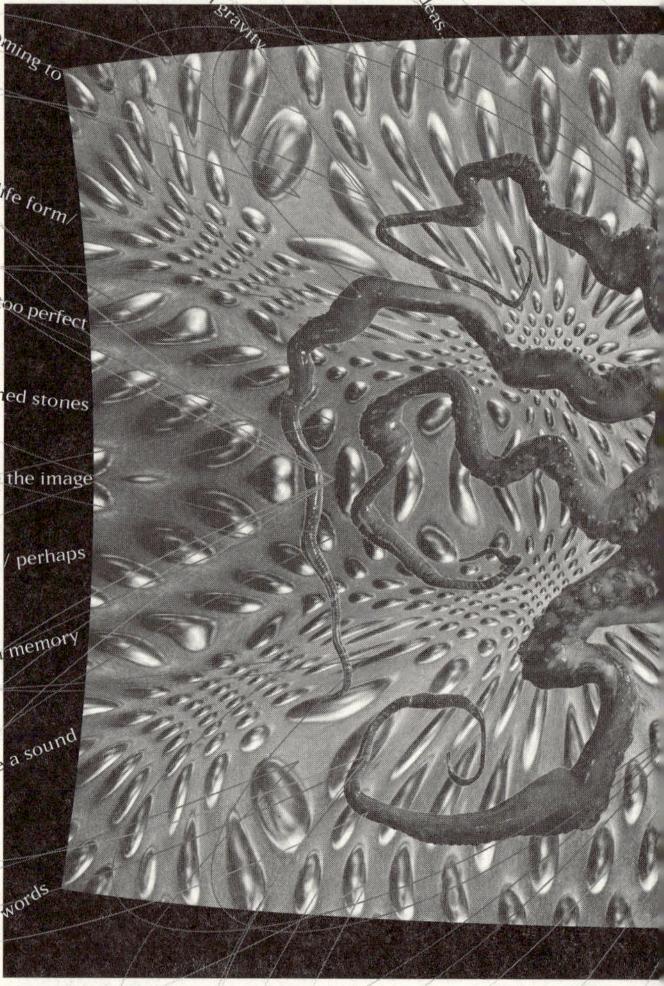

words became flesh

and the whole ocean of life

was funneled through

the great tin horn of language

and language swam upstream

and memory came alive and

kept people awake all night

with its hollow clanking industry...

The eye sees / other eyes / leaping

the eye lights / on other eyes /

leaping / on a whim of a butterfly's wing

It was pouring ⚡. Really pouring ⚡⚡. The rain and wind made visibility through the wire-screened windows almost impossible. It wasn't like the day Jay was released. That was a classic. Then, the sky was a bright blue and sunshine bathed the parking lot as we watched from behind the fence as Jay's hot, little ex-girlfriend drove into view in her little sports car. Jay was all smiles as he left the main building, his duffle bag over his shoulder. He got a big kiss from his ex, lit up a smoke, his first in six months, and then they hopped in the car and drove away. We got a wave goodbye and the show was over. That was four days ago. Now it was raining cats and iguanas. But like Anton said, "Everyday is a good day to get out of jail."

The night before, I'd tried to pack up my worldly goods. It seemed like a lot of stuff when it was all laid out. Mostly books and towels and socks and sweats. I left all the books in the corner library shelf and stuffed everything else into two huge plastic bags. It was ridiculous. It looked like I'd been at summer camp. How did I accumulate so much crap? It was strange to be leaving but I didn't spend much time thinking about it. It was just another day. But very different from the day I arrived.

Transit from the Riverhead facility was an experience straight out of Dante's Inferno. After dinner my name was called and I had to rush to throw my belongings together, roll up my sleeping mat, say a hurried goodbye to Neil and the others and get the hell out of C-31, the so-called "medical observation" tier where I'd been cooped up with thirty other doped-up bozos for the past two weeks.

"I'm keepin' my shoes on tonight, man," said Little Jimmy. "Somebody grabbed at me when I came outta the shower. I don't want no trouble, no sir. I'm sleepin' with my shoes on, 'case I gotta move fast. I'm a little guy—I don't want no trouble. I got to protect myself."

Jimmy landed on our block late one morning just before lunch. He looked more than a little dazed and confused and immediately flopped on his cot against the bars of the day room and slept with the covers over his head. Later that night he told me how he got here. "I was partyin', man, drinkin' gin, and that was FUN... but you know, I can't take hard liquor, I just didn't know what happened. I guess I got into a fight...cops came and hauled me away...I didn't even have a shirt on my back...I don't know what happened, man. I was all fucked up. The doctor here put me on some serious dope, said my blood pressure was gonna give me a stroke...shit! That stuff knocked me out for two days! The doctor says I can't drink no more or I'll have a stroke for sure! You think they'll give me a shirt when I go home?"

Jimmy was a nice guy. He was probably 5'4" and about 110 pounds sweating wet. He was probably the smallest guy in the unit. But he was a tough little black guy with plenty of street smarts. He was only with us a few days. He was with us because this was the "MO" tier: Medical Observation. According to Neil this was the best unit in the whole damn jail because so many of the brothers were under sedation. The population of this tier, C-31, consisted of about 30 guys. There were 15 cells, with one of the 15 serving as a common lavatory for the boys on cots in the day area. The cells were reserved for the guys who were being held pending transport upstate. These guys were headed for some significant time in the system. It was pretty much the same old stuff for these guys: drug charges, petty theft, assault, breaking and entering, grand larceny, possession with intent to sell, etcetera. The guys without their own cells, about 15 of us relegated to steel-framed cots lined up against the wall of floor-to-ceiling bars, were waiting for court dates or already in the process of doing their county sentences. The offenses were pretty similar to the others with the exception that it may have been a "first offense" as opposed to crimes committed by long time felons. Or, they just may have had better lawyers ...or better luck.

When I arrived, on a sunny day in July, I thought I'd be on MO for a day or two at most. I had been sentenced to serve my time at the DWI facility on The Farm in Yapank. Little did I know the machinations of getting from incarceration point A to incarceration point B. In fact, although I was pretty realistic about my fate, my lawyer had left room for hope that I might not have to serve any jail time if he could persuade the judge, himself a pretty notorious alcoholic, that my time was better served on the job than behind bars. Before the official sentencing began my lawyer "conferenced" with the judge but was quickly back in the court room. No sale. The bottom line: I was sentenced to 90 days which meant I'd serve 60 days, minus 10 days "good time" if I didn't get into trouble behind bars, less the long, painful night I'd spent in jail on the night of my accident almost nine months before. Forty-nine days, total. What a way to spend a summer vacation! And the fun was just beginning.

The rumors started a good 48 hours before the event. There was a kind of weird buzz going around amongst my fellow primate inmates. I didn't know what a shakedown meant though it sounded ominous enough. During the course of the days, weeks, and months of incarceration, almost everyone accumulated a considerable amount of "stuff," which could be considered contraband in the eyes of the "CO's" or, correction officers. The presence of drugs or weapons was obviously not something that could be tolerated. Tobacco was also taboo. But aside from that contraband it could mean any damn thing they deemed unacceptable, especially if it made someone feel more comfortable or human in their caged existence.

The reality was that the "street people" brought the "street" with them, wherever they were, "inside" or "out." The most humble and ridiculous items became tradeable commodities inside. Food was "bought 'n' sold" daily, although theoretically, everybody was to get the same allotment. Casettes, batteries, girlie mags, soap, matches, socks, shorts, shirts, sneakers, sheets, toilet paper, spoons, cups, and even mattresses were on the inmate's commodity exchange. Of course, on the "MO" tier, where everybody was under "medical observation," the biggest currency of all was "medication" and the variety was pretty impressive. Occasionally someone would smuggle in some tiny rock of crack or a cigarette, but the day-to-day pharmacopia from the jail house apothecary provided the bulk of serious commerce. At least half of the tier had some serious addictions or mental/emotional problems which were treated, or mistreated, by the medical staff with everything from "mood stabilizers" like Zoloft or Paxil to heavy duty tranquilizers like Thorazine. Of course, even Tylenol was desirable. My personal prescriptions included a three-times-daily dose of Xanax, the 90's equivalent of Valium, which was considerably prized. When the roll call for the dispensing of meds took place, all but the truly ill raced to line up at the end of the tier to get their drugs from the surly attendant. If you missed the roll call you missed out, period. It wasn't long before I became aware of some of the basic techniques for faking the consumption of medication in order to build one's inventory. All medication was to be "swallowed" in the presence of the attendant and the CO, but at least half the population palmed their pills and expertly feigned washing down their medicine with the provided small paper cups of water. And, thus, the "drug trade" continued on the street-within-a-street, behind bars. This all seemed pretty harmless to me and since I truly needed my medication I didn't have the desire to acquire other goods through trading or bartering my prescriptions. Of course the authorities knew what was going on and also knew there wasn't any real solution but they did have the occasional shakedown to put the fear of God back into the inmates' reality.

The rumor was that the next shakedown was coming very soon. It was Thursday when I first heard about it but it came and went and Friday came and went without incident and not much of anything ever happened on the weekends, so I was told. The tier had two so-called "dorm refs." One for the white boys and one for the blacks. Anton was the young mulatto who held court at his end of the block, tried to keep his homeboys in line and dispensed minor favors and chastisements where appropriate. Anthony was Anton's counterpart for the white inmates. His cell served as counseling center and depot for goods. Both guys had leadership qualities, I suppose, although I never really saw them in action. Both of them served as liasons to the CO's presenting prisoner's greivances, concerns, etc. The rumors came to Anton and Anthony by way of the kitchen help who delivered food or filled the large rolling carts to be dispensed by the designated tier man each day. Before I'd arrived there'd been a number of incidents which had begun to look serious to the CO's. One guy was locked down in his cell for a full week for a scuffle he had initiated. Another guy had to be moved to a different tier because of his "anti-social" habit of spitting on people. There had also been a little hanky-panky with contraband. A couple of the boys tried smoking what they thought might be crack and the smoke caught the attention of the wrong guard. So, things were getting a bit tense. But the bust was still a surprise.

Saturday's routine differed from the other days only in the respect that a couple of hours were devoted to cleaning up the cell block. The floors were washed and the cots straightened up while Soul Train dominated the ever-blaring communal TV as the brothers cheered and carried on. Lunch came and went. The daily trip to the "yard," where we got an hour of fresh air out in the broiling sun, came and went, and the day wound down. Dinner was served at 4:30. Then, the card games commenced along with the chess games and general fraternizing or snoozing while the TV game shows blared in the background. It was about 8:45 when the shit hit the fan. It had been a quiet Saturday night until the Storm Troopers arrived. We heard them marching down the hall from the north side of the facility. It sounded like a small army and it was. Led by a short, stocky guy wearing a captain's white dress shirt (and a pair of black leather gloves!), there followed a squad of about twenty CO's. "On your bunks!", was the order of the moment. Everyone scurried to their cot for the all important "count"...the ritual counting of the inmates which occured four times daily 🚶‍♂️🚶‍♂️🚶‍♂️. This "count" was only a prelude to something else. After everyone had been accounted for, each officer personally "investigated" each inmate's possessions. Socks, dirty underwear, sheets, towels, books, letters, toiletpaper, toothbrushs, sweats, flip-flops, boxes of cereal, apples, oranges, cartons of milk, loaves of bread, peanut butter. Every thing and every body was searched for "contraband". A few of the squad dug deep into the nooks and crannys of the cells, tearing apart anything and everything in the process. Ceilings were scoured and cots overturned with a methodical violence which left no one in doubt of what was going on. The final investigation was a strip search of each prisoner who, after striping to the skin, was told to "squat 'n' spread 'em" so that another officer could inspect anal orifices. It was quite a scene: thirty-five naked men having their butts searched for dope, razor blades, or whatever. After this final humiliation, everyone was ordered out of the cell block and into the hall. The grand finale began. The entire tier was tossed upside down and inside out. If there was any personal property intact before, it was destroyed after this hurricane. The Storm Troopers ᛋᛋ had a field day and within a very few minutes the tier looked like it had been savaged by a tribe of rabid gorillas. Not a square inch was left unviolated. The prisoners were marched back in and ordered to clean up the mess and the white-shirted, black-gloved captain, after a vicious oration warning of further actions, headed our way. He gathered his goon squad and marched out of the cell block. Everyone was in a state of shock 😵😵. The dorm refs, who thought they had everything under control (including the CO's), were completely unhinged. How could this happen? But there it was: a nice orderly cell block turned into a garbage dump in a matter of 15 minutes! The shakedown was complete. All we could do now was forensically forage through the wreckage and repair the damage for the rest of our "Saturday Night in the Pokey."

THE END

Out of the blue

The small blue dot in the night sky seemed barely noticeable to the young girl kneeling in the long grass. The infinite blackness was punctuated with a million such luminescent dots. One trifling speck among such a multitude of sparkles had no more significance than any of the others. How dreadfully vast the darkness. How spectacularly wonderful the stars. The young girl, the child, sat quietly amid the tall grass surveying it all. Even the impertinent, tiny blue dot.

She was late. Her mother had told her to be in before dark. What did it matter now? A few minutes longer, more or less. What harm could it do? Just the same, her parents were going to be upset. Not angry of course. She was certain of that. But they would be upset. Her mother especially. She got upset a lot lately. Ever since the accident. Ever since her sister was—. She stopped. She didn't like thinking about it. Anyway, her parents where not going to be happy.

She couldn't help that now.

It was the sky—the wicked, divine sky drawing her in. She sat there among the grass, the weeds, the tiny blue dot and the flowers, gazing up. Before long her attention was drawn to the miniature speck of blue. It wasn't the brilliance of the star that attracted her. It wasn't even the intensely bluish tint that lured her away from a wider observation of the scene. It was rather the way the tiny thing seemed to shimmer, timidly at first, and then, like some faraway beacon, signaling willfully beyond understanding. She giggled a little and turned away, hesitant to surrender to its invocation.

She stood up brusquely, wiping away the blades of grass that had embedded in her knees. Reaching down to pick up her bicycle, she tilted her head and glanced, almost instinctively, upward. A change had taken place. An alarming, distinct, and utterly captivating metamorphosis. The star appeared to be more than twice as large as it had just seconds ago. The bluish color too had turned to a fierce violet hue. And, what before had been a remote and obscure enticement, was now a fully radiating summons.

She stood on the hill, child in the darkness, and didn't turn away. Couldn't. She watched the star dance about in the blackness, leaping and whirling. She watched the radiant spectacle heave, stretch out, devour the stillness of the night. Until it overwhelmed the darkness. Overwhelmed her. Consuming her in a tempest of light, everywhere, all around, inside!

The yoke of gravity that bound her to the earth suddenly yanked away. Her feet lost contact with the ground.

For an instant—there was no earth at all.

by rick blasdell

Just as quickly the blue light vanished. The signal was gone. She stood on the grassy hilltop and stared into the blackness of the night.

But it wasn't gone. It was there—somewhere. She could feel it. She understood it. Somehow. Not like she understood a street sign or a recipe for making cookies or any such thing. She felt it. Like it had been there, always, flashing.

She gripped her bicycle by the handlebars and pulled it out of the tangling weeds. After a prolonged contented breath of night air, she turned and began the familiar journey back down the hill. Back down toward the street. Back down to her mother. To her father, quietly waiting. Back down to her home.

Freebus and Volus

Deep down in the belly of the Crawly-Catchies lived the frolious Freebs. Frivolous to the very bone, constant in every way, they saturated the frimdom and volus until it wriggles no more (wriggling being what most Freebs managed most). Glumping along in a vacuum of consumption, the toddler Freebs functioned in scatalous celibacy late into the night. Wondrous as they were, these recent Freebs could daunt no task nor puncture any bladder without babbling and brooking their institutes of instability or, generally, lacking any compunction to proflerate their fellow master Freebs and Freebuses, they would dazzle their expatriates with shadow stories. This being said, there was no wonder they breathed the night air or trotted their steepled ponies without a shod of care or acrimony. Systematically, steed after steeple relented to their fabled contraptions and dirty inscriptions. Any frolious Freeb would be hard put to ponder any more poetically than was the norm. This was common Freebus knowledge.

The Crawly-Catchies were mocked for their gullibility so, reading their obits and reciting their conscripts was not a challenging task. Nay, the task was arduous but kinetic. Bubbled and babbled, as was their want, they glumped along in deprecating pleasure, withered and wrinkled as they wriggled their wondrous arses in unison over the toad chairs and mush houses. With each rapturous mile they recorded their passions in tablets of mud. Each voluptuous vengeance was registered and remembered and recited and reclaimed at the end of each day. (They would, one day, become notorious for this libelous habit). As they approached the final frimdom, their buggus and volus responded with an exploding, choking cough–this display being obvious enough to constitute it's own network of informational gluttony. All Freebs felt safe upon the delivery of this fractious frobus. Kilkenny and Rambunction, two of the seaters of the seated beast, were among the wildest remnants of the ramparts' feast. For this, they held first come/first service. Both took advantage of this monastic manual of maniacal wit. Paging through the pages and jumping to the last chapter where, as most knew, the answer lay lounging, flapping and spurting until reluctantly noticed. As silly as this seems, no one bothered plucking these morsels from the written word tome tomb. It just wasn't done! Too loose to trek, Mee Casa and Sue Casa, wriggled round and round till their wisdom winced.

(turn the page)

All those holding Crawly-Catchy identity cards could frolic with the frolious Freebs. But Glumping was limited to frimdom and buggus. The Volus fended for themselves. Kilkenny and Rambunction hosted the proceedings, as was the tradition. The belly of the beast had to move two notches down as the Shulacings bore an extra set of eyelets for future expansion. The town Freebs formed frivolous Freebus teams. Each team unzipped their carpet bagettes to fit another underground slave. This was assumed to be of use but too many years later to be measurable. So, buttons on your underwear and two sheets to the windtunnel, the frolious Freebs glumped along in senseless splendor, as they counted the Crawly-Catchies tumbling over the cliff's edge, yelling their team tune as they fell.

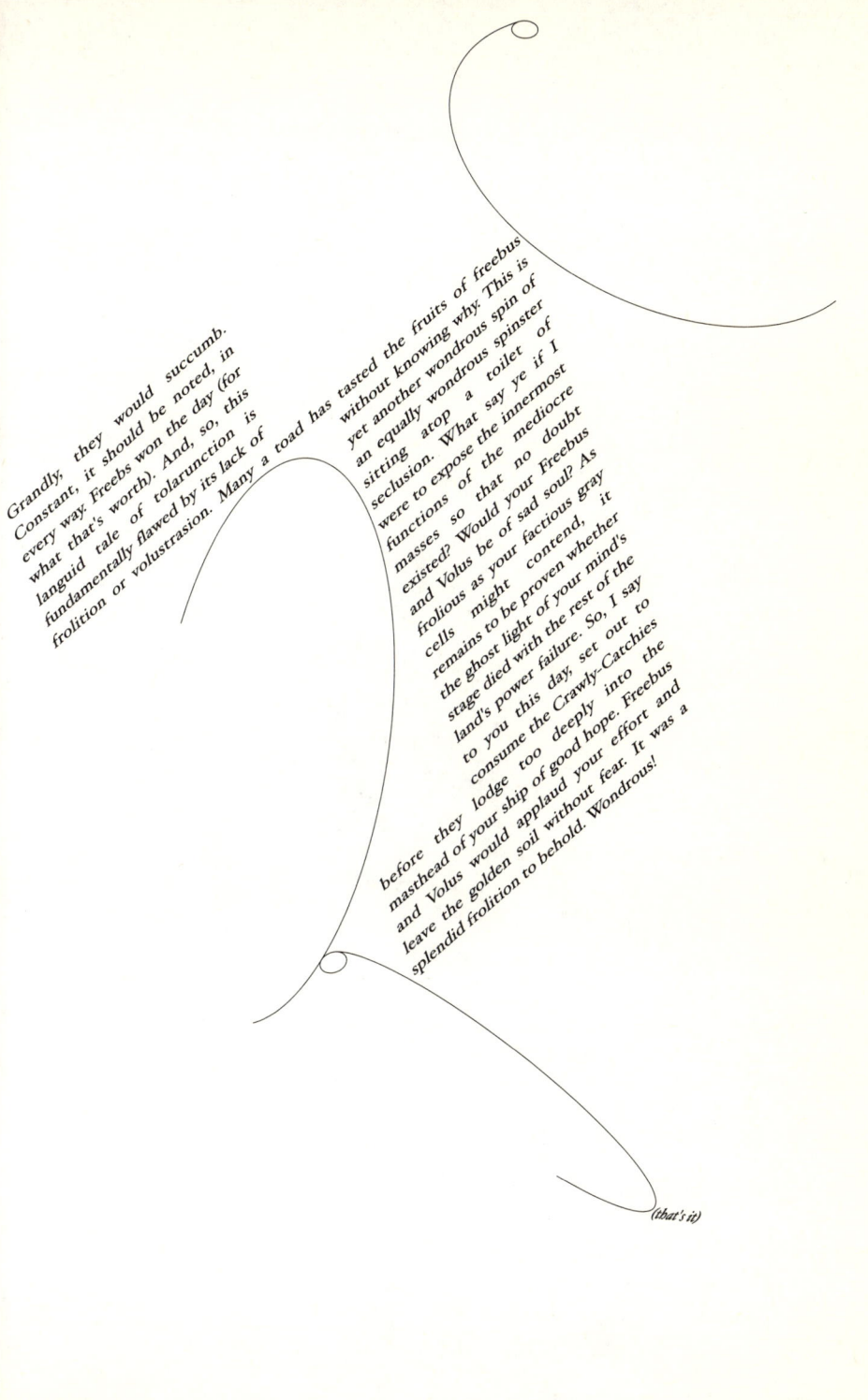

Grandly, they would succumb. Constant, it should be noted, in every way. Freebs won the day (for what that's worth). And, so, this languid tale of tolarunction is fundamentally flawed by its lack of frolition or volustrasion. Many a toad has tasted the fruits of freebus without knowing why. This is yet another wondrous spin of an equally wondrous spinster sitting atop a toilet of seclusion. What say ye if I were to expose the innermost functions of the mediocre masses so that no doubt existed? Would your Freebus and Volus be of sad soul? As frolious as your factious gray cells might contend, it remains to be proven whether the ghost light of your mind's stage died with the rest of the land's power failure. So, I say to you this day, set out to consume the Crawly-Catchies before they lodge too deeply into the masthead of your ship of good hope. Freebus and Volus would applaud your effort and leave the golden soil without fear. It was a splendid frolition to behold. Wondrous!

(that's it)

Nothing holds still /
The seeds of life
are hatched
within the death of stars /
A child grows
& humankind / evolves /
Each child
a new experiment
hatching another scheme /
in the darkness /
twinkling with ideas ...

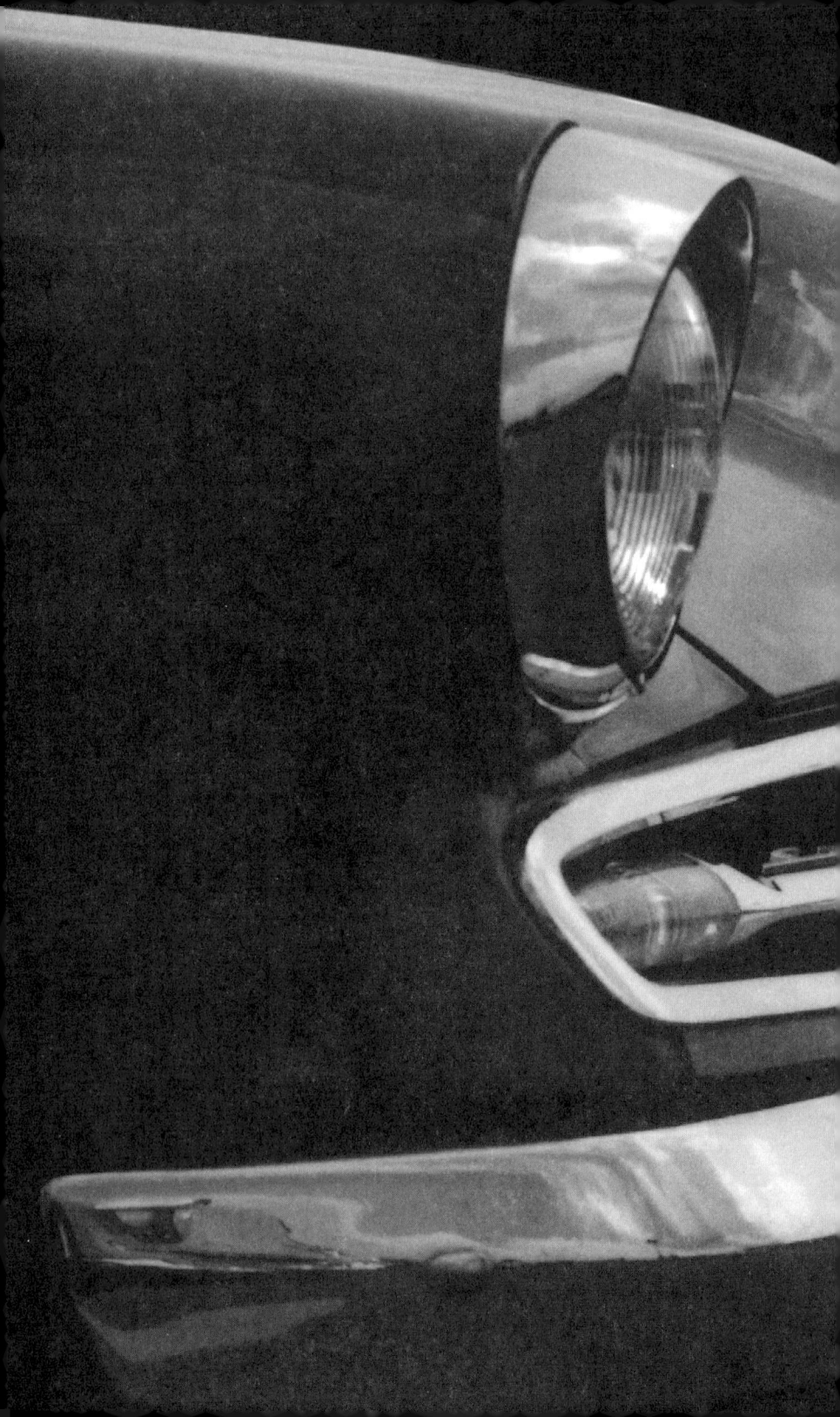

Anthracite Motors presents

Cars from the "Smart Set"

Nº 17 IN A SERIES

#55 the rubber alligator

On Easter morning, March 25, 1959, young Jeffrey Belvedere and his older brother Mike awoke to find two baskets at the foot of their beds. Much to their delight, amongst the chocolate bunnies and jelly bean eggs and marshmallow chicks, each basket contained a ten-inch rubber alligator.

The boys loved their alligators, and proudly displayed them in the family snapshot taken shortly after church service that morning.

Young Jeffrey was especially fond of his alligator. It was always by his side or in his pocket, and after each nightly prayer, he placed it in full view on his nightstand.

One night, just as he was beginning to doze, the alligator spoke to Jeffrey. It said, simply, "Tomorrow, ask your mother for a dozen clothes pins, and use them to attach baseball cards to the forks of your bicycle tires."

The next morning, Jeffrey did as the rubber alligator had instructed and, lo and behold, Jeffrey's bicycle sounded like the radio broadcast of the Indianapolis 500 that he had heard last Memorial Day. After a few spins up and down Fulton Street, he showed all the wide-eyed neighborhood kids the baseball card trick and, from that day forward, Jeffrey was the king of the block.

In the months that followed, the rubber alligator continued to advise Jeffrey on the ways of the world and, in turn, Jeffrey handed down this wisdom to the children of Fulton Street. Despite a burning urge to reveal his secret, Jeffrey told no one about the alligator, not even his brother Mike, who had foolishly banished his rubber alligator to his sock drawer just a few days after Easter.

Later that summer, Jeffrey was playing in his sand box with his favorite neighbor, Claudia Throckmorton. Jeffrey loved Claudia, and felt that she was the only person he could tell his secret to. Placing the rubber alligator in front of her, he leaned forward and whispered, clumsily, "Claudia, I double-dare you to keep a secret. All that I know in this world was told to me by this alligator."

"Yeah, right. And my Barbie told Ken to send Sputnik into outer space."

After a lengthy silence, Jeffrey excused himself and, saddened by Claudia's response, wandered into the house to use the bathroom.

And then, for the first and last time ever, the rubber alligator spoke to someone other than Jeffrey.

"Claudia," said the rubber alligator. "When you are in tenth grade, your trigonometry teacher, Mr. Blevins, will give you an A minus on your report card just because he daydreams about what color underwear you have on."

Hearing a scream, Jeffrey glanced outside the bathroom window just in time to see Claudia dashing across the street towards her mother's house.

Three days later, a for sale sign appeared on the Throckmorton's lawn. Each time Jeffrey knocked on their back door, Claudia's mother politely told him that she was doing homework. Two weeks later, the Throckmortons moved to Nebraska.

Following the incident with Claudia, the rubber alligator spoke to Jeffrey less and less, sometimes in languages that sounded to Jeffrey like something out of a Tarzan movie. On November 12, 1959, after a month of silence, the rubber alligator said, simply, "I am very tired." He never spoke to Jeffrey again.

Before school on the morning of April 13, 1960, Jeffrey buried the rubber alligator in the tomato patch next to the sand box.

Y'tink y'know whatcher want
CAUSE Y'VE COME TO LIKE
whatcher already got.

The tv sells you
over & over
the same
whatcher already got.
Laughter cuts
th'thin blue air//
skeletons on th'rooftops
electric blades
that carve a night
out of pulsating hieroglyphics/
A pixilated/grotesquery
with its face prest against
my tv screen//

We have standards for excluding people
& ARE FAR TOO BUSY TO CARE//

Can we bleed it out with leeches??? Can we leech it out with jesus???

The self-organization of complex systems:

Stable enough t'transmit information

yet lively enough t'mutate & evolve.

I sometimes wish I was younger and

had this much on th'ball

cause at th'age I am

right now

THIS just doesn't seem like

all THAT much to have ...

on th'ball ...
Between what is
& what should be
these monsters dance
on my tv/

Information will be feeding US

We'll grow up on knowledge
Knowledge will be power//the arrogant
juice of expertise
can be suckt from th'root of a cable/
Even the sun is irrelevant now/

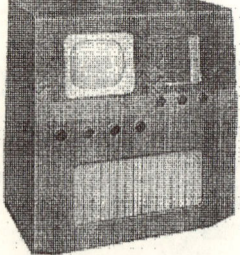

Alien satellites circle th'globe
comin home t'roost ...
th'face of a monster
hangs in my window/
th'face of a monster
on my tv

devouring his vague notion of a piece of the pie//

> His empty fifteen minits/
> thumpin in his brains//
> th'secrets of th'universe
> rocket thru his veins/

**RICOCHET ... CHAOTIC ...
SEARCHIN FER TH'PERFECT PRODUCT//**

The geometry of atoms
smashing their way thru a brutal ballet
stamping vicious message craters
into each other skulls/
tiny universes
ricochet off of each other/

SEPARATE

but wiser
How can there be
television
where there is no
vision at all?
waddaya tell a vision

thadda vision
should be tellin you?

Excerpts From The Izzy Tapes

by gene hosey

To people who ask me "why?"
I like to ask back "why bother?"
After all
you can never end up being
whatever it is you might imagine
someone else imagines you to be /

We are not the things we say /
the things we say're what's left of us:
smoke on the end of a burning rope /
smoke
that makes us dare to think
that someone else'll think of us
when were not with em in the room /

Some people consider their hobbies
their Zen
and some mistake Zen for their work
and say pithy things like
"The only thing a man truly owns
is his work"
and
"Gravity is the motion of a coin
to the bottom of a pocket"...
but work / work is / something
you do for others /
for money /
dressed in their clothes /
miserable business
the ones with the money
either aren't willing or
aren't able to do themselves.

The pictures referred to in "The Izzy Tapes" can be found in the book "Une Semaine De Bonté" [A Week of Kindness] by Max Ernst

This very day I sat
in my fact'ry parking lot
toastin wee crawly ants
with my bifocals.
I used to burn ants as a kid
under the sun thru a magnifying glass /
and I can't figure out
how the feck
I ever got to be this old.

Today was Tuesday. Bats and serpents
crawled over everything. People
were marched naked in chains by
wealthy un-uniformed citizens.
I did my part. I was
distracted by the fact that
I could hear the dragon claws
etching on my office door.
But on Tuesday I often
sprout barbed wings myself /
I have bills
that must be paid
and debts I owe myself.
And I understand the dragons
who hold the chains
along these streets they've built
we live on.
They trust me to speak for them /
they threaten me constantly /
but otherwise don't pay me
much attention /
so I've gotta describe
these pictures to you:
I live among the chained beasts
along the streets
the dragons prowl /
like a winged dragon m'self
buried alive in this bone box
scratchin at my lid of flesh
clawing my own chest
trying to breathe
normally
inside an abnormal whirled.

Deep within the 1950's
when things seemed so much
Quieter
the writers were at work
busy with lighting the
thousands of fuses that
blew the big bang 1960's
strait from the heart
of the dense black subatomic
invisible spec
of my childhood.
The revolution was ON
and I was coming to it late
more willing than able
locked and slightly loaded.
"People in groups
groups with names
uber-people /
like-minded people who gather to sandbag
who wanna make noise
but don't wanna be seen /
ignoring everything that works
dropping what's in hand
to line up
to jump in
behind the next new thing /
motivation / inspiration / perspiration
needs analysis options design:
these shiny new chains
we wear/ to make us look smart
Tell me:
what ever happened to
"I don't feckin know?"/

We are lucky people /
we are from
different whirleds
you n'I
whirleds
obviously communicating with each other
though / somewhere / messages are getting
garbled /
We are lucky people:
we could be ridin around in pick-up trucks
frownin / with our teeth knocked out /

Ach/
these chest pains
again today / maybe my heart
finally givin way /
More likely though
corrupted intestines /
those feckers are
wrapped around
everything /
But tell me now:
where's the glory in
kickin off from a
bitched up colon?
Not t'mention prolonged
agonizing pain /
Ah
knock me down
with something sudden /
drag me off the field /
But this? This is a whirled of
penance
and you must EARN your end of penance /
feck the pain
git back t'werk /
keep movin /

I'm not sure if
this woman is my wife
or just another worker /
she
is either pleading
or reporting
to the bald man at the desk /
she stands there rigidly
straight backed / securely dressed
buttoned up tightly ankles to neck
hair done up like bunches of grapes /
She complains to the man at the desk
there's too much rain / too little
no getting no work done today
no use tryin to plow /
I sit and
rub my forehead
comb my fingers thru my hair /
thru the windows
all the same:

too much rain / too little ...
But for those of us who can listen
we persevere / we smile / we wave
and we find that
it's easy to hate everybody
yet / still be successful socially /
nuthin makes y'feel better
faster / than runnin into someone
feelin just as bad as you /
it's what
gets me outta bed
mornins /
What would be worth getting up for
if life as it is was as good as it gets?

"What" the bald man asks the woman
"is practical about being human being?
Aside from buildings and bridges
cars and starships
and our ability to blow everything up /
what is there that's practical
about a human being?"
"Jeeesus!" she sez /
"does everyone really
have to have
a point of view
on everything?"
The bald man / in a giant bird's head
now stands and puffs a pipe:
"Yeah, our species did all dat:
we put a man on the moon
Mars / further / we're spending a
gigajillion dollars
to launch a 2-foot wide
exploring plane
that'll rocket for 48 hours
across the face of a red planet
snappin photos / steadily being crushed
and baked in the thick red atmosphere
and crashing
alone into doom / we can do dat!"
His explanations
so mechanical / so slapped down
dressed up
trotted out /

and y'ask the next guy
and he tells you the same thing
in the same sequence
and it's like
everyone's seen the same teevee show
but *you* fell asleep
and now there's yet another thing
that you know nothin about /
Since when
do we need a conspiracy
t'lead ourselves astray?

"Oh sure," she sez
"I know all about
what an arrogant clever species
we are / and what about the rest of us?
Will we get to go to Mars?
Sure / yeah sure
we're all whether we want to or not
gonna be goin to Mars /
Where d'ya think
we'll be getting the slaves
to mine the metals
weld the ships / and retrieve the lil
red robot planes?
Y'think there'll be unions
on spaceships and Mars?"

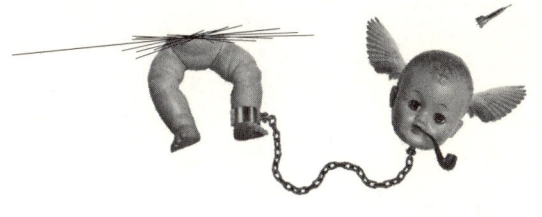

"The humans, by definition, need to assess the universe's general qualities; it's how they anticipate, to some extent manipulate, and so survive in it. ❷ To generalize, they look for patterns: mother holds me and feeds me if they cry, things fall down, soil and seed and water make food, etc. ❸ Simple or sophisticated, depending on previous experiences, but all useful stuff.

"❹ They discover certain tricks can help: for example, characteristics common to several patterns form a meta-pattern which binds together essential knowledge; and they're even better off if they associate this meta-pattern with frequently reinforced experiences of daily life, since ease of recall is crucial. ❺ So, the more familiar, emotional, intimate, or deeply felt the experience, the better it serves. ❻ A broad, multi-functional shape emerges: father-sky waters-inseminates mother-earth, who holds-draws-feeds us.

"❼ They keep it by repeating it and using it; they maintain it lovingly. ❽ Myth, narrative, story, whatever -- ❾ I like to call it a tool, which it is. ❿ Sequentially. ⓫ Note, though, that even if stuff lumps neatly our strings well ⓬ it doesn't necessarily imply affinity or cause-and-effect relation beyond the narrative. ⓭ There's no such thing as 'fertility' infusing both womb and soil; it's an abstracted characteristic, developed from the observation of functional similarity. ⓮ Likewise, the law of gravity rightly deals ⓯ earth's maternal image a lethal blow.

"⓰ Still, you might reasonably characterize 'the human' as 'one who studies nature's ebroader physical properties', 'seeking resemblances to its own ... [illegible]

"Now, after several thousand years of these experiments, the humans' toolbox approaches consummate differentiation. A question arises: having fashioned such awesomely funky gear, can the human contrive the Right Job for the Tool?

"The answer--that the human fears and barely understands its tools--is perhaps an accident of evolution. After all, radically new artworks or self-critical philosophy are impossible luxuries when life is short,

memory defective, writing nonexistent, and survival crucially knowledge-
  dependent.
  Human
     social
       evolution
         depends
           on
             effective
               effects--freezes
                 it--and
                   on
                     cosmologically
                       meaningful
                         the
                           pace
                             at
                               a
                                 form
                                   of
                                     mythic
                                       application
                                         and
                                           celebratory
                                             rite.
  Or sacrifice, whether the development of new tools depends on wide-ranging
  freedoms; but the hominid tinker--to tinker with the arts
  mythologies, or technologies--to not fear change--courted suicide.

"However, even if this species spends the rest of its days documenting its own decay, it will never diminish this one sublime accomplishment: collaborative self-portrait, unique contribution to the cosmos, the species' raison d'etre.

"We the Tool--We the Grapheme--say: Bravo.

"Curtain call. The creature bows, stands, nurses its tender shoulder; swings the tool with a broad, tentative stroke. Yes, it's beginning to feel sure and familiar--the new limb, faculty, or sense is emerging. It swings again. Exhilarated, it gazes into the white lamps and considers the angle. A double flash of bare metal and the creature is gone--having routed its sockets, deftly, onto the proscenium. Where the Tool also lies.

"Thanks, it's been real. We'll take it from here. Don't forget to sign our species' guestbook, folks."

Typographica
by d. clark

Anthracite Motors presents

Cars from the "Smart Set"

Nº 45 IN A SERIES

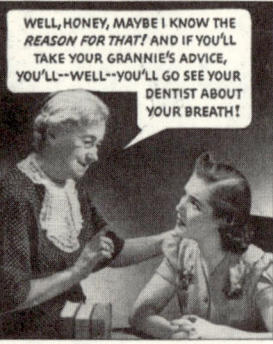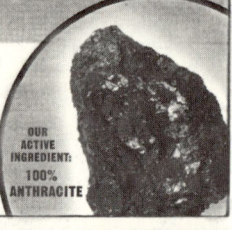

HE Mailed This Coupon

E. W. HOSEY
Anthracite Champion
Cup Winner

This is an ordinary snapshot of one of Charles Anthracite's Pennsylvanian pupils.

...and Here's the Handsome Prize-Winning Body I Gave Him!

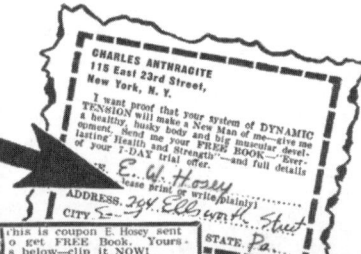

E. W. HOSEY saw my coupon. He clipped and mailed it. He got my free book and followed my instructions. He became a New Man. NOW read what he says:

"Look at me NOW! 'Dynamic Tension' WORKS! I'm proud of the natural, easy way you have made me a Champion!"
—E. W. Hosey

"I'll prove that YOU, too, can be a NEW MAN"— *Charles Anthracite*

I don't care how old or young you are, or how ashamed of your present physical condition you may be. If you can simply raise your arm and flex it I can add SOLID MUSCLE to your biceps—yes, on each arm—in doublequick time! Only 15 minutes a day—right in your own home—is all the time I ask of you! And there's no cost if I fail.

I can broaden your shoulders, strengthen your back, develop your whole muscular system INSIDE and OUTSIDE; I can add inches to your chest, give you a vise-like grip, make those legs of yours lithe and powerful. I can shoot new strength into your old backbone, exercise those inner organs, help you cram your body so full of pep, vigor and red-blooded vitality that you won't feel there's even "standing room" left for weakness and that lazy feeling! Before I get through with you I'll have your whole frame "measured" to a nice, new, beautiful suit of muscle!

Only 15 Minutes a Day

"*Dynamic Tension!*" That's the ticket! The identical *natural* method that I myself developed to change my body from the scrawny, skinny-chested weakling I was at 17 to my present super-man physique! Thousands of other fellows are becoming marvelous physical specimens—my way." I give you no gadgets or contraptions to fool with. You learn to develop your strength through "*Dynamic Tension*." You simply utilize the DORMANT muscle-power in your own God-given body—watch it increase and multiply double-quick into real, solid LIVE MUSCLE.

My method—"*Dynamic Tension*"—will turn the trick for you. No theory—every exercise is practical. And, man, so easy! Spend only 15 minutes a day in your own home. From the very start you'll be using my method of "*Dynamic Tension*" almost unconsciously every minute of the day—walking, bending, etc.—to BUILD MUSCLE and VITALITY.

FREE BOOK "EVERLASTING HEALTH AND STRENGTH"

In it I talk to you in straight-from-the-shoulder language. Packed with inspirational pictures of myself and pupils—fellows who became NEW MEN in strength, my way. Let me show you what I helped THEM do. See what I can do for YOU! For a real thrill, send for this book today. AT ONCE.
CHARLES ANTHRACITE, Dept. 8ZB, 115 E. 23rd St., New York, N.Y.

CHARLES ANTHRACITE

An untouched photo of Charles, the winner and holder of the title "The World's Most Perfectly Developed Man."

CHARLES ANTHRACITE
Dept. 8ZB, 115 East 23rd Street, New York, N. Y.

I want the proof that your system of "*Dynamic Tension*" can help make me a New Man—give me a healthy, husky body and big muscle development. Send me your FREE book, "Everlasting Health and Strength." No obligation.

Name..
(Please print or write plainly)

Address......................................

City.....................................State...............

In four years of marriage you never wanted to talk like this.

I have trouble looking in your eyes when I try to tell you things. That's about you, not me. But go a h e a d

This is new ground...it's hard to s t a r t

Your turf here...I'm so desperate to hear you talk about anything that I'll play along at this point.

I guess I'm worried and frustrated and completely bored. Mostly scared, really.

Go on

It's pathetic, but the cliché has come true. Work is nothing but corporate masturbation. Now they want me to fire four supervisors. Found that out today. They masturbate away some people's lives and go out of the room all sweaty and sticky and slapping themselves on the back. They're so fucking small... they fucking think small and they act small.

What's happened at that place, anyway? You used to be so gung-ho about it.

There's a crazy reverse physics going on – the inertia of not being able to start anything good, the momentum of greed.

What can you do about it?

Nothing. Not a thing. And I sit and think of how incredibly unimportant this all is in the scheme of things, how a detail like this gets lost in the billions of fucked-up things that happen every day. And I figure, why care anymore? So I sit in my office and get lost in my head. I can't work any more. I can't focus or care about any of the shit going on there. And it scares me that I can't work.

What are you going to do?

I just stay and it makes me sick. I'm one of those fucks who is just holding on. I'm the guy I made fun of 10 years ago.

This goes beyond work...you're the same at home. You don't look in my eyes when we're fucking anymore, either. Are you even in the room when we're having sex? Or are you far away, like when you're in your office?

I'm there, probably more than you realize.

But how is it that you can fuck me and drink me down and make me come but you can't look into my eyes or talk to me? Why are you so giving in bed and so closed off the rest of the time?

I don't know. Maybe I want to please you, in some way at least. Maybe to make up for what I don't give you.

You don't know what I want. That is so arrogant. Why are you so convinced that the things you think are so right?

Are you telling me you don't want more from me? I'm a stranger. I hate that.

It doesn't seem to affect your life much. Not enough to change, anyway. You like the luxury of not joining in. It's so weird...you're ravenous when we have sex, but hardly even there the rest of the time.

What do you want me to say? That I'm afraid to show how completely fucking limited I am? Or that I know exactly how I'm ruining our relationship, but can't make myself do anything about it? And I mean I know exactly how, down to every single hurtful detail.

That's your interpretation. It's in your head. I never said you're at f a u l t.

I do something wrong. I say something wrong. Or I don't say something. I can see it, like I'm floating above watching the entire episode unfold. And I know just what I've done, and how I could undo it. But I can't m o v e.

What holds you back?

I feel so lousy for making the mistake that it just kills me. I feel the weight of all the mistakes piling up and I can't breath. Every time it happens I feel more clumsy and stupid.

But why is it so hard to undo? Is it the way I am?

No, it has nothing to do with you. That kills me, too. You pay for no reason other than I'm clumsy and paralyzed and stupid.

You do know that everybody is fucked up, right? It's pretty arrogant to think of yourself as so superior that you can't make a mistake.

I don't want to be like everybody else. I have the capacity to be better.

But you're not doing better. You're doing worse. Because all those imperfect beings out there can undo it...

...tell me something, have you ever felt good enough about anything you've done? Including being with me?

I feel good about being with you. Not as good about you being with me...

...why are you looking at me like t h a t?

I appreciate this...usually, when we try to talk, you turn it into an intellectual game or twist my words around to fit your agenda.

I feel good about it, too. So what are you going to do tomorrow? What are we going to d o?

Get up. Go to the office and figure things out. Come home, take a walk with you. Get on with t h i n g s.

Are we going to have more conversations like this?

Y e s.

Any thing else you want to t e l l m e?

Just that I'm really sorry I've stayed away for so long.

Is this the way you hoped this conversation would go?

Better, way better.

Written by Bill Spinner, a freelance writer in Harrisburg. He's written travel and historial pieces for The Patriot-News and is a frequent contributor to Harrisburg Magazine. Bill is the screenwriter of The Last Covered Bridges of Pennsylvania, a feature documentary that aired on the statewide PBS network in 2000. An avid traveler, his ambition is to live in Rome and write travel pieces on his laptop from the sunshine of Piazza Navona.

one
hundred
and
sixty
four
people
that
i
would
like
to
have
a
beer
with

do
you
know
them
all?
anthracite.tv
does.

Designer as Amusing Chatchke

Designer as
Stylish
Vase

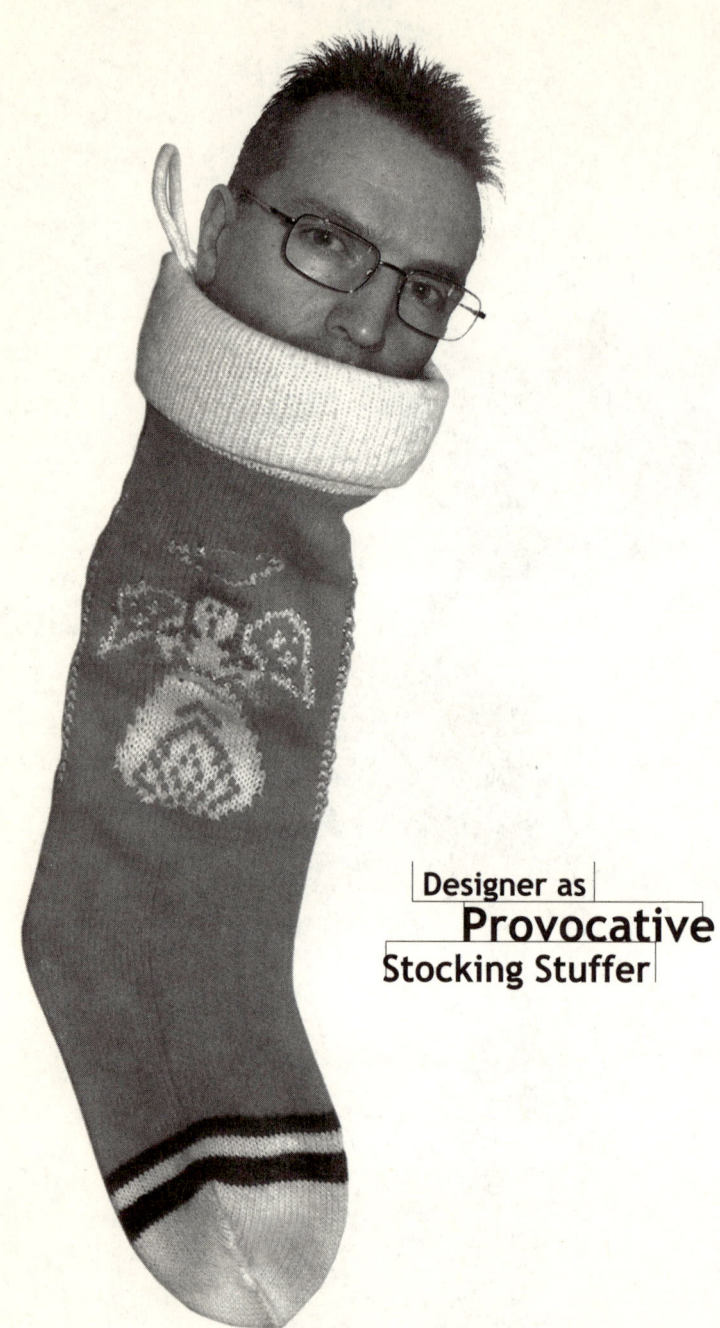

Designer as **Provocative** Stocking Stuffer

Designer as National Icon

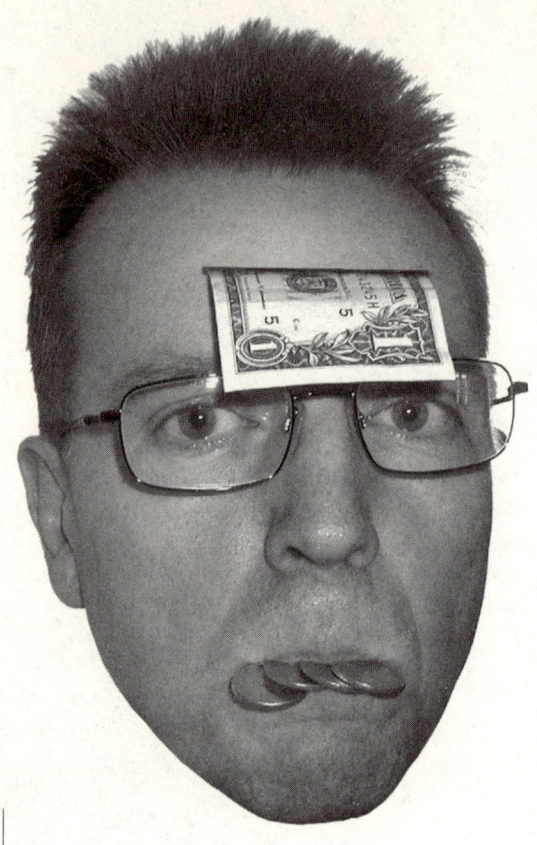

Designer as
Convenient
Money Changer

Designer as
Charming
Finger Puppets

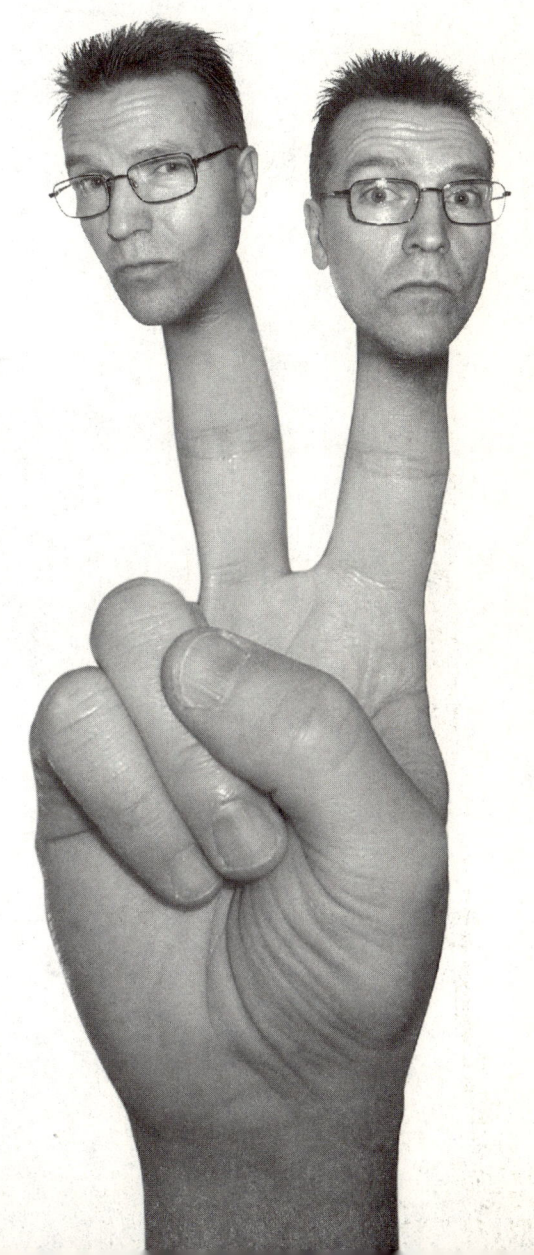

THE BURNING HEART

sketches by
RICK BLASDELL

The twelve illustrations printed here are enlarged, retouched versions of a series of random, very small, concept sketches.

The original drawings are about the size of a postage stamp.

story by
ROBERT HUMBLE

Copied on individual sheets of paper, in no particular order, the illustrations were handed to Bob. As a challenge he was asked to construct a story from these drawings, assembling the pages any way he would like.

What he wrote, in about a fifteen minute period and nearly word for word, is what is printed here.

I shall tell you a tale of love. I will tell it to you so that it has a happy ending.

For tales of love, no matter how tragic, are happy.

I found him on a field of glory, amidst the horror he seemed serene.

And on him I found a letter—

I knew at once the author's response when she would hear of this great battle—whether glory was gained or lost by the hosts that met here.

For her, the pain would remain.

The letter was written in a delicate hand.

The hand of a girl, an educated girl—a lady? —A princess?

She gave it to him as he rode away— for evermore.

Honor and duty called him away.

For he was noble and true to his word. A word he gave solemnly.

How could she resist what she knew was right?

For had she not also sworn an oath to be pure and wait for a sign?

A sign to fall upon the one true love that would be hers forevermore.

Her youth had been spent in self-absorption, she knew that her gifts were many.

Yet a gift is worthless unless given to another. A lesson she had not yet learned.

The boredom was great before he came. She could find no meaning in life.

Her eyes were closed to all that surrounded her.

And then one day the noble youths found each other.

Perhaps at a banquet—

perhaps in a temple—

or —was it in a glade?

They met and gave meaning to each others lives.

For a time, how brief? They loved—and the wonder of the world was clear.

And this dear readers is the happy ending.

For though it may not have lasted—what ever does?

Love, honor, truth— it is only they that last forever.

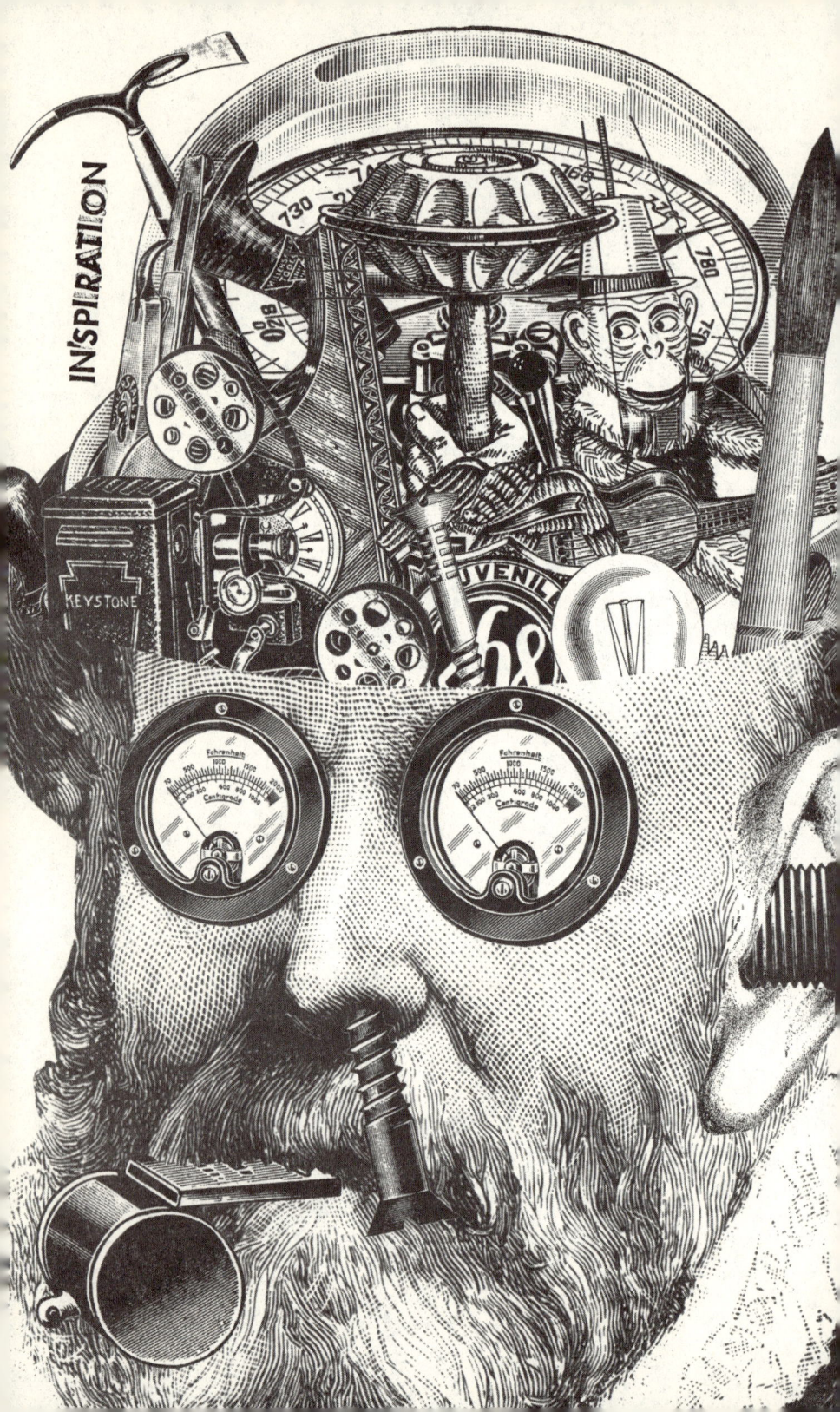

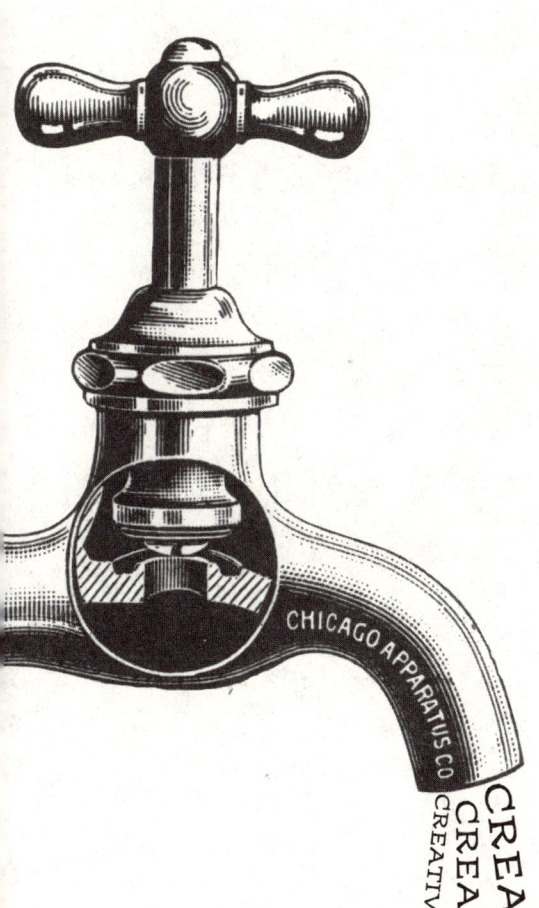

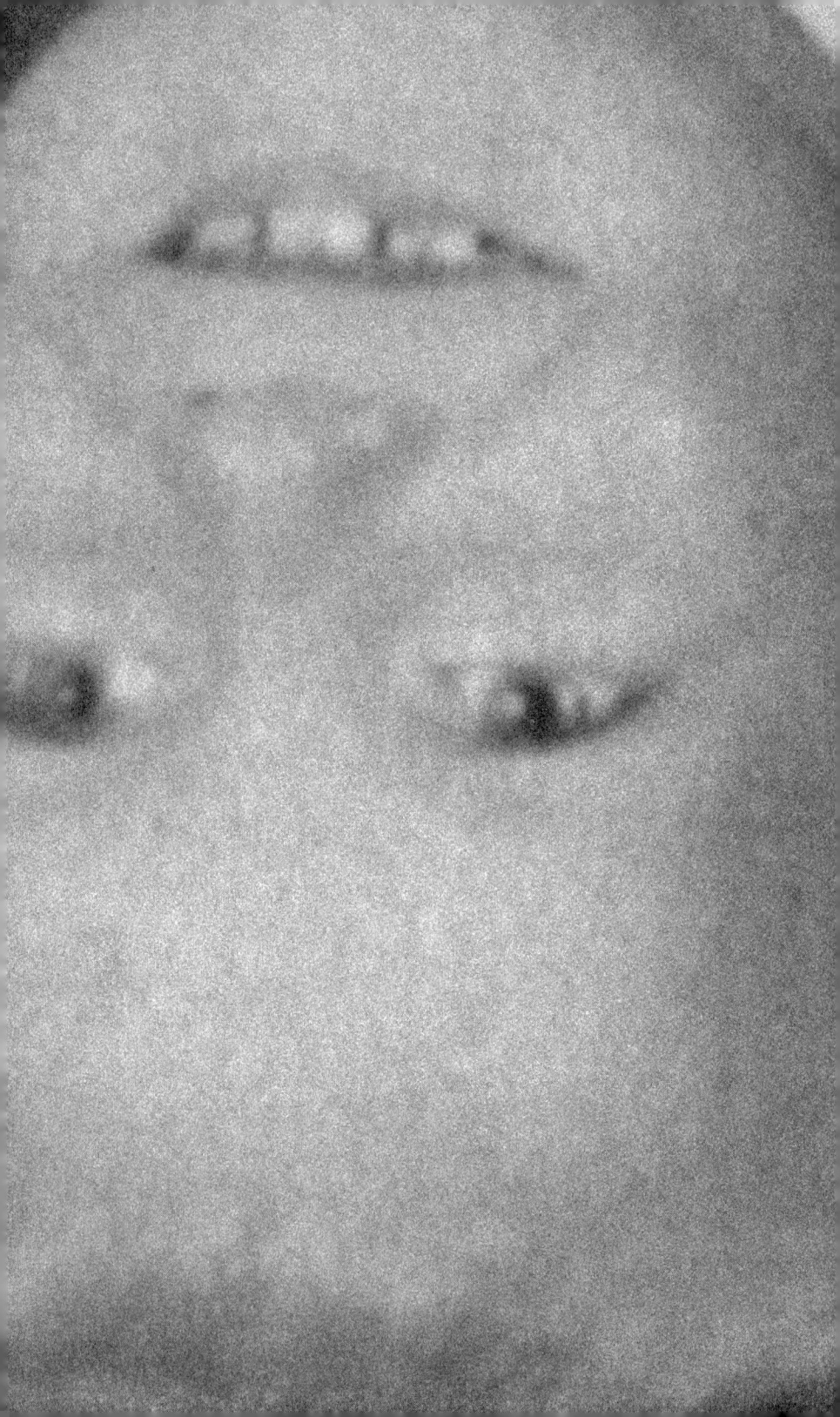

I'm a graphic designer. I create things. I bring visual order to the life around you. People with no talent of their own, but with alot of money to spend, hire me to make you think their way, buy what they produce, and do things you wouldn't ordinarily do. Mostly, I'm misunderstood and underutilized. The world is a better place simply by my existence. I hold the secret.

Within the context of all worldly things, I'm completely irrelevant. I place words and images onto pieces of paper and become agitated when it's not taken for art. I'm self-centered, self-serving, self-righteous, self-destructive, and self-conscious. My saving grace is that I'm a self-starter and self-contained. My only regret is that I'm not self-polinating. I'm poorly traveled and under-educated. Popular culture is my religion and the computer my messiah. I take full responsiblity for my actions so long as I ultimately get my own way. I speak of my work as if it were original, useful and effective but scoff at the work of those more talented. Every moment of my existence is spent in search of an ultimate spiritual peace I know I'll never find.